Forest Certification Auditing

SECOND EDITION

Forest Certification AUDITING

A GUIDE FOR PRACTITIONERS

SECOND EDITION

William Rockwell and Charles Levesque

Society of American Foresters
Bethesda, Maryland

Published by
The Society of American Foresters
5400 Grosvenor Lane
Bethesda, MD 20814-2198
www.safnet.org
Tel: (301) 897-8720
Fax: (301) 897-3690
0-939970-95-3

Copyright © 2005, 2007 by the Society of American Foresters.

All rights reserved. No part of this book may be reproduced in any form or by any means, electronic or mechanical, including photocopying, recording, or by any information storage and retrieval system, without written permission from the publisher. Address inquiries to Permissions, Society of American Foresters, 5400 Grosvenor Lane, Bethesda, MD 20814-2198.

Acknowledgments
The authors thank John Hodges, who provided the basis for several of the FSC case studies, and Lynn Penniman, who had valuable suggestions on auditor selection. We also thank Lynn Penniman and Kathy Abusow for their reviews on the first edition, and Guillaume Gignac and Dan Simonds for their reviews of the second edition.

Library of Congress Preassigned Control Number 2007938036.

Contents

Overview vii
1. Background of Forest Certification Auditing 1
    Origins of forest certification 3
    Dimensions of forest certification schemes 6
2. Major North American Forest Certification Schemes 8
    Canadian Standards Association 8
    Forest Stewardship Council 9
    Green Tag Forestry Certification 9
    ISO 14001 10
    Sustainable Forestry Initiative® 11
    Tree Farm System 11
3. Auditing Principles, Roles, and Procedures 13
    Participants in a forest certification audit 13
    Sequence and schedule for an audit 15
    Auditor qualifications 15
    Auditor conduct and responsibilities 17
    Preliminary document review 18
    Audit plan 19
    Audit scope 19
    Audit objectives 20
    Audit criteria 21
    Preliminary audit 22
    Field audit 23
    Collection of evidence 24
    Findings and conclusions 26
    Reporting 28
    Nonconformance remedies 29
    Appeals 30
4. Forest Certification Case Studies 31
    SFI Case Studies 32
    Tree Farm Case Studies 39
    FSC Case Studies 43
    Some answers to forest certification case study questions 47
5. Resources 54
    Getting started as an auditor 54
    Forest certification websites 55

Overview

This book originated in an 8-hour course that the authors originally provided to their subcontractor auditors employed on forest certification audits carried out by their firm, The Plum Line. The course was augmented and modified for presentation to more general audiences, including both people interested in becoming forest auditors and those interested in becoming certified. In conjunction with the Society of American Foresters, the course was offered at the SAF convention in Winston-Salem, North Carolina, in October 2002; at Jacksonville, Florida, in July 2003; and at the October SAF conventions in: Buffalo, New York, in 2003; Fort Worth, Texas, in 2005; and in Pittsburgh, Pennsylvania, in 2006.

The course has been well received, and there were some requests for course materials by people unable to attend the course itself. In addition, SAF is helping its members prepare for its Certified Forest Auditor® exam. This book is intended to serve both purposes.

The book provides an overview of universal auditing principles applied to the unique characteristics of forests and their management, independent of any particular forest certification scheme. As such, it is meant to supplement other courses and materials aimed at general environmental management system auditing, surveys of forest certification schemes, and auditor training for forest certification schemes (all of which also are important for forest auditors and, to some extent, for firms seeking to be audited). By filling this special niche, this book is intended to bridge the gaps among forest certification schemes, raise the quality of all forest certification auditing, and provide a basis for improving forest certification standards. It is not intended, however, to take the place of forest certification auditing training required for any auditor certification or other formal auditor qualification.

Chapter 1 concerns forest certification auditing in general and covers the theoretical and historical origins of the "green" certification of forests. Chapter 2 describes the characteristics of major North American forest certification schemes—Canadian Standards Association, Forest Stewardship Council, Green Tag, International Organization of Standardization, Sustainable Forestry Initiative®, and Tree Farm System. Chapter 3 covers auditing principles, roles, and procedures applied in the context of forests and forest management. Chapter 4 presents 15 case studies to demonstrate the application of the material in Chapter 3. Chapter 5 provides an overview of other information resources on forest certification, including how to get started as an auditor.

1 Background of Forest Certification Auditing

Throughout history and in virtually all cultures, legitimate public interests in forests have been recognized. Forests have been, at least historically, a source of strategic—generally military—materials and have provided essential raw materials for economic development. The industries they support have been a basis for maintaining economic stability by providing jobs, and the abundance of wood products has enabled many people to afford decent housing. In this way, as well as through the provision of public recreational access and facilities, forests have contributed to social and economic equity.

Forests provide ecological as well as economic services. Forest cover is the preferred land use for ensuring the purity of water, and forested watersheds release this high-quality water more gradually than do agricultural fields and developed land. Forests provide habitat for fish and wildlife, offer aesthetic and recreational resources, moderate temperatures and climate, and promote air quality. Such services—water, wildlife, aesthetics, climatic and atmospheric effects—are "public goods" that move across property boundaries.

To secure all of those important interests, governments have supported direct production in public forests but also have promoted forest management for private forests through subsidies, technical assistance to landowners, and regulation. These tools have not always been effective for several reasons, however, including difficulties in reconciling multiple forest goals and values and the complexity of and variation in forest systems. As an example, Figure 1 shows a general model of the complexity of a forest system with multiple goals; a more detailed model would create links that would be impossible to follow.

Although the complexity of forests should be obvious; the variation among forested systems is often overlooked. Not only are forests very different under different geological and climatic conditions, they also vary greatly within a given forest, through the year, and from year to year. In addition, forests are not contained in pipes or factories but spread out on the land, and they are non-engineered systems—there are no blueprints telling us how they work. Although forests are stationary, the public goods they generate are mobile. All these factors mean that forests are costly to monitor, and their status and condition are subject to significant uncertainty. The forestry profession is one potential source of such information, but because foresters serve landowners (and not the public at large), there is no assurance that foresters will be motivated to produce public goods that are neither desired by the landowner nor

2 FOREST CERTIFICATION AUDITING

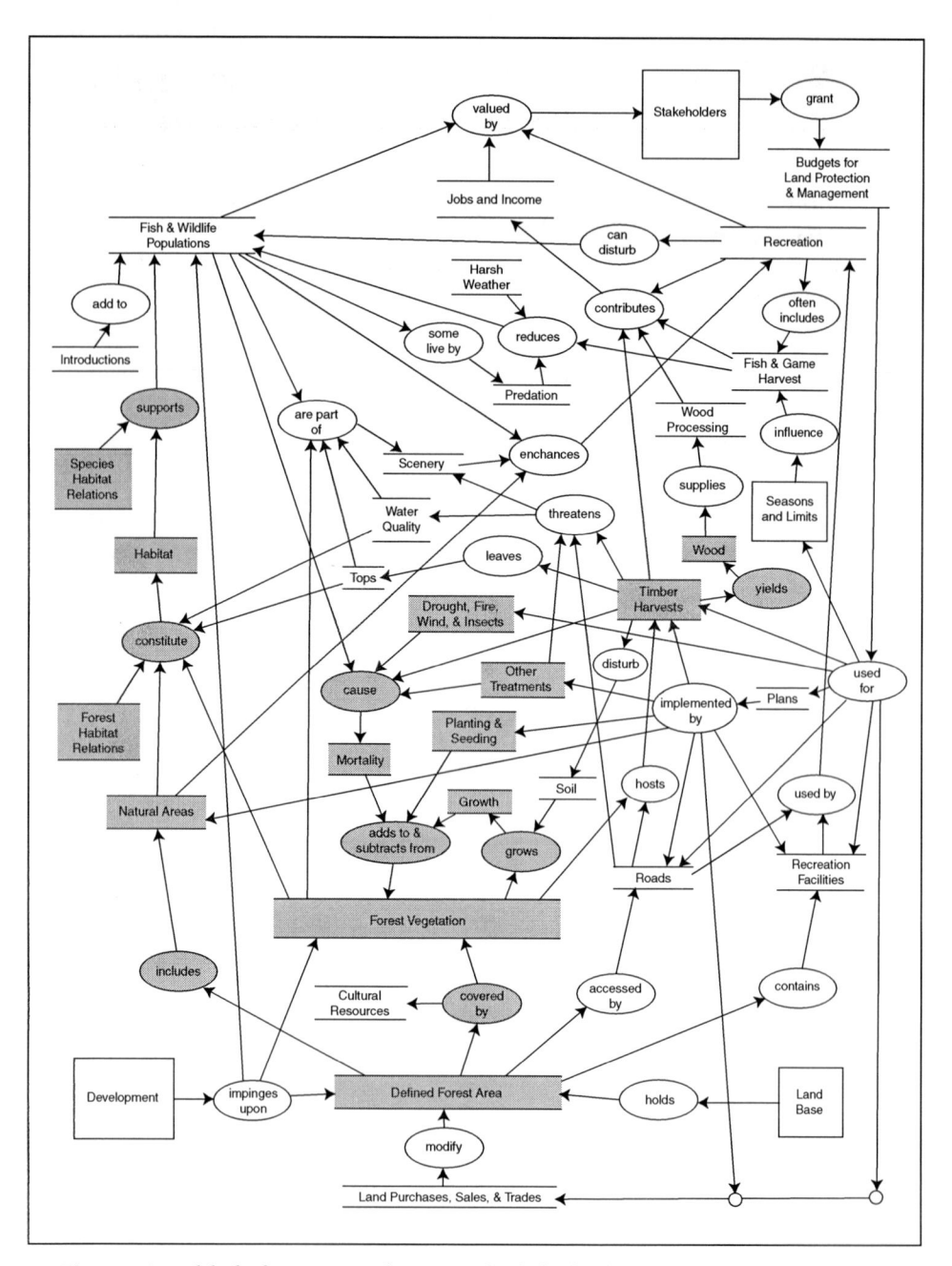

Figure 1. A model of a forest system showing multiple feedback loops.

required by law. As a result, the public has turned to independent entities that can reliably report on forest conditions (see Figure 2).

Because of the need for trust and the failure of traditional mechanisms to reliably ensure environmental performance, private certification schemes have sprung up in recent decades to handle a wide range of concerns. The International Organization for Standardization (ISO), for example, offers certification to its 14001 environmental management system (EMS) standard, which addresses the impacts of companies and other types of organizations on the physical environment. In the ISO's "plan–do–check–act" framework, an organization evaluates and develops goals, objectives, targets, plans, and operational procedures for managing its significant environmental impacts. It then monitors the results and makes adjustments to continually improve its environmental performance.

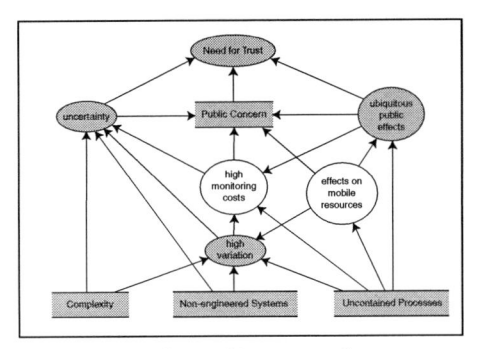

Figure 2. Relationships among forest characteristics leading to a need for trust.

But complex forest systems do not just *affect* the environment—they *are* the environment. In addition, there is no "do–nothing" alternative against which the effects of management can be judged, and management is intended to have desirable effects, not just avoid undesirable effects. Guide ISO/TR 14061 was developed to describe how an ISO 14001 EMS can be applied to the complexity of a forest using a relevant set of national or multinational criteria and indicators listing potential concerns about forests. Significantly, the uniqueness of forestry makes ISO/TR 14061 the only sector-specific guide developed under the ISO EMS approach, which otherwise prides itself on a level of generality that transcends industry-specific concerns.

Still, IS0 14001 leaves much to the discretion of the organization seeking certification, so it cannot guarantee that specific performance standards will be met or even that particular public concerns will be addressed. To more reliably ensure specific forest conditions and outcomes, therefore, forest certification schemes have arisen that more directly address the public's wide range of environmental, economic, and social concerns about forests.

Origins of forest certification

Although numerous forest certification schemes have blossomed in the past decade, such schemes are not new. The first forest certification scheme, the American Tree Farm System, was established by the US timber industry in

1941. Notably, Tree Farm was an alternative to federal regulation of private forest practices, proposals for which had long been resisted by private landowners. The traditional Tree Farm program involves a certification inspection by a certified inspector, who also may give technical advice. Recently revamped to meet international certification standards, Tree Farm has added a group-certification option (discussed below). In combination with other voluntary conservation programs, some states' forest practice regulation, and the natural regrowth of forests that had been logged, the Tree Farm program succeeded in addressing public concerns about forestry for many years.

In the 1980s, the general public became aware of the extent of tropical deforestation, and these concerns led to demands that the developed countries stop importing tropical timber. The International Tropical Timber Organization pointed out, however, that tropical deforestation (which is driven more by permanent land-clearing for grazing and agriculture than by timber harvesting per se) would actually be encouraged by trade embargoes that made sustainable forestry less valuable than alternative land uses. The key to promoting a more effective approach to slowing deforestation would be giving wood-importing countries the ability to distinguish wood from well-managed forests from wood coming from indiscriminant forest clearing. As forest certification schemes emerged to meet this need, a private organization, the Forest Stewardship Council (FSC), was formed in 1993 to provide consistency in certification standards and auditing protocols and to accredit certification bodies.

Governments also were considering ways to address the issue of tropical deforestation, among a host of other environmental issues. Not surprisingly, they emphasized the importance of government-to-government relations recognizing the primacy of national sovereignty, as well as the balancing of environmental protection with responsible economic development. The landmark 1992 "Earth Summit" (formally the United Nations Conference on Environment and Development, UNCED) took a major step forward in this approach with its "Agenda 21" (the Rio Declaration on Environment and Development), including its Chapter 11, "Combating Deforestation," and its Annex III, "Statement of Principle for a Global Consensus on the Management, Conservation, and Sustainable Development of All Types of Forests." This work led directly to the founding of regional protocols for assessing the state of forests according to an agreed-on set of criteria and indicators issued in the Santiago Declaration in early 1995: the Montréal Process applies to non-European temperate and boreal forests; the Helsinki Process, to Europe; the Tarapoto Process, to the Amazon; and several others, to Africa, the Near East, and Central America.

Although these processes focused on the national level and did not include targets or forest-level assessment methods, they reinforced the concept that objective measurement of forest management quality, as was then being promoted by FSC, was desirable. By emphasizing national sovereignty and action, moreover, these processes helped spawn the development of nationally defined, forest-level assessment and—eventually—certification schemes, in part in reaction to FSC's claim to transnational legitimacy as a nongovernmental organization (NGO).

In the United States, meanwhile, more states had been establishing forest practice regulations. Responding to growing pressure for expanded and stricter regulation, the American Forest & Paper Association in 1994 developed an industry code of conduct, the Sustainable Forestry Initiative® (SFI). Five years later, in part to provide a nationally appropriate alternative to FSC certification, SFI was transformed into a third-party certification scheme, and in 2004 the standards were revised under the auspices of the now-independent (since 2001) Sustainable Forestry Initiative, Inc.

Also in 1994, the ISO (the international organization of national standards bodies), long a developer of international technical standards, devised a standard for assessing the claims of organizations regarding their quality management systems (ISO 9001). Then, as organizations began to make claims for their environmental quality management systems, a new standard, ISO 14001, was developed in 1996 on the pattern of ISO 9001 to assess and certify these environmental claims. In the same year, drawing on the Montréal Process criteria and indicators (C&I), the Canadian Standards Association developed its explicitly ISO-based, C&I-oriented Sustainable Forestry Management Standard, CAN/CSA Z-809. And in 1998, the National Forestry Association established the Green Tag certification scheme to serve primarily non industrial private forests.

To provide for efficient certification of small landowners, many national schemes have developed group certification schemes, in which several to many landowners under a common management system can be certified if sample forests are found in conformance without the certification auditors needing to inspect them all. These arrangements are most notable under FSC, Tree Farm, and many of the European schemes. In addition, a new certification scheme for small private woodlands, CSA Z804, is currently under development in Canada.

Finally, to provide for a mutual recognition scheme for national certification schemes (and as an alternative to that provided for its members by FSC), the Pan-European Forest Certification scheme was founded in 1999 and has since developed into the current global Programme for the Endorsement of Forest Certification (PEFC).

Dimensions of forest certification schemes

Forest certification schemes vary along several dimensions that reflect their origins and the strategies they developed to serve the purposes that they set for themselves. These dimensions include the following:

- Scheme and standard development processes;
- Content of standards;
- Scheme scale and scope;
- Mandatory or voluntary nature;
- First-, second-, or third-party claims;
- Certifier accreditation processes; and
- Mutual recognition framework.

These dimensions, as defined below, will be applied to six major North American forest certification schemes in Chapter 2.

Scheme and standard development processes. Schemes and standards are developed by different sponsors, with varying degrees of openness and with the involvement of different kinds of participants and stakeholders.

Content of standards. Standards include different ranges of considerations, involve different degrees of specificity, and emphasize different elements and different values.

Scheme scale and scope. Certification schemes vary by scale; although all schemes consider their standards applicable at any scale, more detailed standards apply more readily to larger organizations. They also vary by the type of forestry organizations, processes, and products they address. Schemes may certify forests, forest management, forest managers, forest management systems, wood procurement systems, wood products, product "chains of custody," and finished wood-based products (some of which may include nonwood materials).

Mandatory or voluntary nature. Schemes vary in the degree—or, more accurately, the conditions under which—participation and certification are mandatory or voluntary.

First-, second-, or third-party claims. Schemes differ in the options formally available for the public use of audit claims. Standard audit industry language differentiates "first-party" (self-declaration of conformance), "second-party" (attestation by a separate party that may have a direct interest in the auditee's operations), and "third-party" ("independent") claims, and this language is used throughout this book. The term "certification" (also called "registration") is generally reserved for the declaration of conformance resulting from only a "third-party audit". It should be noted, however, that the industry-standard practical requirement for "independence" allows the auditee to select and pay its own auditor, so auditors should be prepared to address the question often posed by others of "who is the 'third party'?" in a "third-party audit".

Certifier accreditation processes. All schemes require some recognition of the certification bodies (or registrars) that audit to their scheme, mandate certain credentials of individual auditors (especially lead auditors), and have some procedural, reporting, and audit requirements for certification bodies.

Mutual recognition framework. Different schemes offer or use different approaches to recognizing the results of other schemes' audits as equivalent to their own.

"Performance" vs. "systems" standards. One final aspect of standards that is widely discussed is the degree to which various forest certification schemes embody an "outcome" or "performance" approach as opposed to a "process" or "systems" approach. It is supposed that, in the former, performance is judged on its own merits regardless of plans; in the latter, it is said, forest managers must have good plans but are not evaluated on whether those plans are actually carried out to good effect. The polarity of this difference is typically overstated, however, as performance is the test of the effectiveness of a system, and performance cannot be judged independent of the goals and methods of the system that produced it. In addition, this dichotomy often is confused with the unrelated dimension of the degree of prescriptiveness of a standard with respect to its required process or outcome elements. Finally, in forest certification auditing, the intent, implementation, and results of a forest management system all matter and must fit together, regardless of the manner, clarity, and consistency with which these aspects are formally embodied in a standard. Because of these complexities and subtleties, which will be addressed in more detail in Chapter 3, no attempt will be made to characterize forest certification schemes according to a performance versus systems approach in Chapter 2.

2 Major North American Forest Certification Schemes

The characterization and classification of certification schemes require simplification, and simplification invites controversy. It is not the intent of this book to provide a full history, analysis, or comparison of forest certification schemes, but a brief discussion of North American schemes along the dimensions discussed in Chapter 1 can provide insights to assist in their application. No comparative judgment is implied by this comparison.

The following schemes are described in this chapter (in alphabetical order):
- Canadian Standards Association (CSA);
- Forest Stewardship Council (FSC);
- Green Tag Forestry Certification;
- ISO 14001;
- Sustainable Forestry Initiative (SFI®); and
- Tree Farm System.

Although all potentially important in the eventual success of forest certification schemes, three developments related to forest certification will not be covered by this analysis: forester certification, "point of harvest" certification (such as logger certification), and forest product and "chain of custody" certification labeling, although the last of these will be mentioned under "scheme scale and scope", below.

Canadian Standards Association. CSA's Sustainable Forest Management (SFM) scheme was first published in 1996 as the CAN/CSA Z809 standard, which was subsequently revised in 2002.

- *Scheme and standard development processes:* The scheme was developed in a multistakeholder, consensus, and open process by primarily governments, academics, scientists, aboriginal representatives, consumer groups, industry, and some environmental NGOs through the Canadian Standards Association (CSA); other NGOs opted out of the process.
- *Content of standard:* The forestry standard is comprehensive and encompasses a substantial public involvement component, a requirement to address 6 national criteria and 17 forest-level elements of performance, and a management system component modeled on ISO 14001.
- *Scheme scale and scope:* The CSA scheme is more suited for large ownerships, especially on Crown/Public lands with their existing public involvement

requirements. The forestry standard is augmented by an optional product label and chain-of-custody requirements.
- *Mandatory or voluntary nature:* CSA is a voluntary scheme.
- *First-, second-, or third-party claims:* It accommodates third-party certification audits only.
- *Certifier accreditation processes:* Certifying bodies must be accredited by the Standards Council of Canada, and auditors must be certified as EMS auditors; ISO procedures are required for both and conform to PLUS1133 for certification bodies and PLUS 1134 for auditors.
- *Mutual recognition framework:* CSA received the endorsement of the Programme for the Endorsement of Forest Certification (PEFC) in March 2005.

Forest Stewardship Council. Founded in 1993, FSC was originally an accreditation body for independent forest certifiers and their standards.

- *Scheme and standard development processes:* This global organization has established 10 principles and 56 criteria for national and regional certification bodies and their standards (which add more local indicators to the standard principles and criteria).
- *Content of standards:* The principles and criteria are supplemented by many local indicators (e.g., 163 for the Central Hardwoods–Lake States standard). Emphasis is on the balance of three "chambers": environmental, social, and economic, and explicitly requires public input on forest management audits.
- *Scheme scale and scope:* FSC has been adopted more on larger ownerships, but new adjustments are available for small and low-intensity managed forests. It certifies forests, landowner groups, foresters, chain-of-custody systems, and labeled products.
- *Mandatory or voluntary nature:* The scheme is voluntary.
- *First-, second-, or third-party claims:* FSC offers third-party certification only.
- *Certifier accreditation processes:* FSC authorizes the certification bodies, which then recognize and govern their own auditors. Under a new policy, auditors are required to meet ISO 19011 qualification requirements by the end of 2005 (FSC-STD-20-004, qualifications for FSC certification body auditors).
- *Mutual recognition framework:* FSC was created as a mutual recognition framework, and does not recognize nor is recognized by any other framework.

Green Tag Forestry Certification. Established in 1996, Green Tag is an independent forest certification scheme for smaller, private landowners.

- *Scheme and standard development processes:* Green Tag was developed by the National Forestry Association in cooperation with consulting foresters and the National Woodland Owners Association.

- *Content of standards:* The scheme involves 10 criteria and 46 indicators, emphasizing forestry processes and outcomes.
- *Scheme scale and scope:* This scheme was designed for smaller private ownerships; it covers both forest certification and product labeling.
- *Mandatory or voluntary nature:* Green Tag is a voluntary scheme.
- *First-, second-, or third-party claims:* The scheme provides only third-party certification. Green Tag is unique among the third-party schemes described here because its auditors are selected and paid by the National Forestry Association rather than by the auditee.
- *Certifier accreditation processes:* Auditors are recognized, selected, and guided by the National Forestry Association.
- *Mutual recognition framework:* None.

ISO 14001. The ISO first published its international, generic EMS standard in 1996 and revised it in 2004. Although not a forestry standard per se, and often maligned for forestry use because it does not set its own explicit forestry performance criteria, an ISO EMS can provide a rigorous basis for developing a forest management system that addresses any formal certification standard.

- *Scheme and standard development processes:* ISO 14001 is offered under the auspices of the International Organization for Standardization, the global association of national standard-setting bodies.
- *Content of standards:* ISO 14001 consists of an outline of required elements for an environmental management system, augmented by several supporting standards (e.g., for life-cycle analysis and labeling). Its forestry content is suggested by ISO TR 14061; auditing and auditor qualifications are guided by ISO 19011:2002 (Guidelines for Quality and/or Environmental Management Systems Auditing).
- *Scheme scale and scope:* The scheme can be scaled up or down but is more appropriate for larger organizations. It covers management systems, but the generic character of ISO facilitates integration of certified wood into products containing other materials as well.
- *Mandatory or voluntary nature:* ISO 14001 is voluntary; it confers some regulatory relief in certain jurisdictions.
- *First-, second-, or third-party claims:* All three are possible, but only third-party audits justify a claim of certification—"registration" in ISO terminology.
- *Certifier accreditation processes:* In the ISO scheme, certification bodies must be vetted by national accreditation organizations, and individual auditors are approved by certification bodies following international requirements and guidance.
- *Mutual recognition framework:* ISO accreditors follow international protocols under the auspices of the International Accreditation Forum (IAF).

MAJOR NORTH AMERICAN FOREST CERTIFICATION SCHEMES

Sustainable Forestry Initiative®. SFI originated in 1994 as an industry code of practice established by the American Forest & Paper Association (AF&PA), which added a certification component in 1999. The standards were revised in 2000, 2001, 2002, and, most recently, 2004; the latest standards are intended to be effective through 2009.

- *Scheme and standard development processes:* The 2005–2009 standards were developed through an open public process by the now-independent governing organization, the SFI, Inc.
- *Content of standards:* The standards set forth 13 objectives, 33 performance measures, and 102 indicators covering forest management, utilization, wood procurement, some social factors, and the support of research and professional and public education; a supplement, Audit Procedures and Qualifications, requires auditing procedures to follow ISO 19011.
- *Scheme scale and scope:* The SFI standards are aimed at larger forest holdings and procurement systems. In scope, they cover forest management, wood procurement, and processing, with optional labeling and chain-of-custody standards.
- *Mandatory or voluntary nature:* Scheme participation is mandatory for AF&PA members and also available on a voluntary basis to nonmembers; certification is optional for all scheme participants.
- *First-, second-, or third-party claims:* All three options are available. First-party annual reports to the SFI, Inc. are required, but only third-party audits can result in public certification claims.
- *Certifier accreditation processes:* Certification bodies must have had ISO-14001 registrar status as of 2006, with a special SFI-registrar accreditation process available in 2007.
- *Mutual recognition framework:* The SFI program received the endorsement of the Programme for the Endorsement of Forest Certification (PEFC) in December 2005, which carries with it the requirement to conform to the international protocols of the International Accreditation Forum. The SFI labeling standard recognizes the equivalence of wood from Tree Farm sources.

Tree Farm System. The Tree Farm System was developed in 1941 by US industry as a scheme for forestland owners; recently upgraded, it now includes a group certification process.

- *Scheme and standard development processes:* The Tree Farm standards were developed by the American Forest Foundation.
- *Content of standards:* Each of the 9 standards has a few supporting performance measures and indicators focused on forest management; more

extensive documentation supports group certification.
- *Scheme scale and scope:* This scheme is aimed at smaller landowners or groups of small landowners, and it covers forest operations only.
- *Mandatory or voluntary nature:* Tree Farm is a voluntary scheme.
- *First-, second-, or third-party claims:* Both second- and third-party audits are available to individual holdings, but Tree Farm groups must undergo third-party audits.
- *Certifier accreditation processes:* Tree Farm trains and certifies its auditors and group certification bodies.
- *Mutual recognition framework:* Its certification is recognized for SFI procurement systems through the SFI on-product labeling standard. Tree Farm is currently aspiring to recognition by the Programme for the Endorsement of Forest Certification.

3 Auditing Principles, Roles, and Procedures

This chapter considers the general principles of auditing applied to the unique context of forest certification, including the roles of various actors in an audit, the auditing procedures that are used, and the typical outcomes of the process. These topics can be covered in a general way only, as each forest certification scheme and audit firm have their own auditing policies and procedures. The treatment in this chapter is intended to provide a common basis for understanding the individual cases of the various schemes and firms.

Participants in a forest certification audit

The actors in an audit need clearly defined roles that prevent confusion about their individual rights and responsibilities. The roles are established in the written procedures of the certification scheme and/or the certification body, and are laid out for each person by name in the audit plan. Typical roles include the following:

- *Client:* the organization—or more specifically, the client's representative—that is commissioning the audit. "The client may be the auditee, or any other organization with a regulatory or contractual right to request an audit" (note to 3.6 of ISO19011:2002, "Guidelines for Quality and/or Environmental Management Systems Auditing") and must be at least willing to participate in the audit—even if not enthusiastically.
- *Auditee:* the organization to be audited. The auditee and the client may be the same or, for example, a large corporation may be the client, and the auditee may be one of its divisions or regions. The auditee is represented by a specific person overall; additional representatives of the organization who have specific geographic or functional roles, such as the field manager for a forest district, also may have specific coordination roles in the audit.
- *Certification body, audit firm, or registrar:* the organization conducting the audit (and, usually, issuing any certificate).
- *Audit manager:* the representative of the certification body in charge of the audit.
- *Lead auditor:* the individual leading the audit on the ground.
- *Audit team:* auditors who report to the lead auditor. Team members often have training or experience in conducting audits and usually have specific areas of technical expertise. They may be responsible for evaluating certain topics, such as wildlife ecology or forest engineering, and they often make decisions in their fields.

Some small or annual maintenance (or "surveillance") audits can be conducted by the lead auditor alone if that individual has all the technical qualifications needed for the audit; even in such audits, however, a second auditor provides a second set of eyes and ears, a sounding board, and support for an auditor who might be facing a defensive or uncooperative auditee. Audits to more than one standard may have a different lead auditor for each (especially if more than one audit firm is involved), even though they typically share the other members of the audit team.

Some certification bodies use an audit manager. Technically, the audit manager is generally not a member of the audit team per se; this person handles contracting and billing, often arranges the logistics, and when needed, is the auditee's appeals contact for the audit firm. The audit manager chooses and contracts with the lead auditor who, as team leader, typically helps select and always supervises the other team members. Either the audit manager or the lead auditor (or both) might be responsible for delivering any additional training deemed necessary for team members to fulfill their assigned audit roles.

Some audits may have several other actors:

- *Sponsor:* a client, such as a buyer or the holder of a conservation easement, who commissions an audit (with the auditee's permission) but has no more than a limited direct interest in the auditee. For example, a foundation may provide a grant for a state lands audit.
- *Assistant lead auditor:* an auditor who may represent the lead auditor in his absence, as when an audit team splits up to cover dispersed sites.
- *Technical expert:* a member of the audit team who cannot audit but advises the team members about technical matters associated with the audit.
- *Auditor (or lead auditor) trainee:* a technical expert or otherwise inexperienced audit team member who works under sufficient supervision to build qualifications to perform in the future as an auditor or lead auditor.
- *Observers:* representatives from the client or sponsor organization (perhaps including their certification consultants) or the auditee's stakeholders. Their limits in the process should be obvious, but some observers may have to be reminded.
- *Certification scheme, accreditation, or certification body representatives:* participants involved in routine oversight; interpretation of audit standards or procedures; documentation of audit results; appeals of audit conduct or findings; or in some schemes, the issuance of certificates.

Appropriate and specific rights and responsibilities for each role should be laid out in the audit plan, reviewed in the opening meeting, and reinforced as needed during the audit. The lead auditor is typically responsible for all these tasks (as well as working with the auditee to ensure the safety of audit participants).

Sequence and schedule for an audit

After contracting and selection of the lead auditor, a third-party certification audit generally proceeds as follows:

- Selection of the audit team;
- Preliminary remote review of the auditee's documents;
- Preparation and submission of the audit plan;
- An onsite "scoping," "preliminary assessment," or "readiness review";
- A formal determination of readiness, often with conditions for proceeding with the audit;
- A period of weeks to months in which the auditee responds to any deficiencies found in the preliminary review;
- The initial certification audit;
- An audit report (sometimes in preliminary form, with the remedies for any nonconformances), which normally includes a recommendation regarding certification; and
- The audit firm's issuance of the final audit report and subsequent award of a certificate, if warranted.

Auditor qualifications

Audit firms and auditors must maintain both their own reputation for credibility and that of their forest certification scheme. Credibility is protected through a range of objective criteria; a set of important personal attributes; and accreditation, certification, and monitoring systems to ensure both.

Essential personal attributes for auditors include the following:

- Technical knowledge appropriate to the subject matter of the audit;
- Understanding of the forest certification standard and scheme being applied;
- Forest certification auditing skills;
- Communication skills;
- Organizational skills;
- Fairness and impartiality; and
- Reasonableness, discretion, and good judgment.

Evidence of those attributes can be gained from records of technical and auditing training and experience; the testimony of former auditees and those who audit the auditors; written, systematic audit procedures (including those related to auditors' qualifications); documented claims of independence and the lack of conflicts of interest or bias for or against the auditee; formal or informal references and recommendations; and formal audit firm accreditations and auditor certifications based on these factors.

Collectively, a forest management audit team must know forest and wildlife ecology, local forest ecosystems, forest hydrology, forestry and related environmental regulations, and practical forest management systems and operations. Each member of an audit team, however, need not possess all the knowledge, auditing skills, and the personal attributes listed above in equal measure. Subject matter assignments are generally made according to major sections of the audit standard. Auditors are generally chosen for their individual technical knowledge in a combination that most efficiently provides the range of expertise needed on an audit team, although auditors who possess broad knowledge or enough auditing experience and skill to work alone have an advantage. A lead auditor, on the other hand, must know the forest certification standard and scheme thoroughly and possess excellent auditing and organizational skills and personal qualities; specific technical knowledge is less crucial as long as other team members can close that gap. Note that individual forest certification schemes and audit firms have their own auditor qualification requirements, which may include participation in scheme-specific training sessions.

To ensure that qualifications are assessed objectively, auditors and audit firms must honestly represent their professional qualifications in résumés and proposals, during self-introductions in an opening meeting, and in explaining and justifying their determinations in the audit findings. In the final analysis, admitting to a limitation in one's qualifications is less damaging to one's reputation than the possible discovery of an unwarranted audit finding; honesty can only enhance an auditor's credibility.

The auditor selection process generally begins when a client seeks proposals from one or more audit firms. Requests for proposals vary in their complexity and formality, but typically, an audit firm submits the résumé of at least the proposed lead auditor—and perhaps the entire audit team—with a proposed audit schedule and protocol and an estimate or bid of the associated cost. If the proposal is accepted, the lead auditor then participates in or leads the selection of other audit team members drawn from a formal or informal pool of qualified auditors. Questions about any actual or apparent conflict of interest between each auditor and the auditee—including recent employment, contracting or consulting, other financial interests, and family or other close personal relationships—are essential; every prospective audit team member must make full disclosure and give assurances of the lack of such conflicts.

Because an auditee is naturally inclined to select an audit firm based on cost, convenience, and even the anticipated ease of passing an audit, audit firms and managers must often resist the temptation to compromise on audit team skills, experience, conflicts of interest, or time commitment. Formal audit-firm procedures for defining such qualifications and commitments help bolster this resistance, but all accredited schemes also require external monitoring by the accreditation body as part of the certification body's accreditation requirements to ensure that such procedures—as well as the certification scheme's other requirements for audit objectivity—are followed. Conversely, clients and auditees must realize that their selection of auditors reflects on the public credibility of their audit and, eventually, the credibility of the scheme whose certification they are seeking. Time will tell whether the existing system of incentives and safeguards are adequate to protect the objectivity and credibility of the forest certification auditing industry.

Auditor conduct and responsibilities

Auditor confidentiality regarding the audit begins with the initial contact from the audit firm or lead auditor. Terms of auditor confidentiality should be included in all agreements between the audit firm and audit team members, the contract with the auditee, and the audit plan.

Cordiality and respect are to be expected in all audit communications. Forest managers are typically proud of their performance in a complex and unique situation that they know much better than their auditors, but auditees also can get overconfident, develop "blind spots," and practice wishful thinking about the quality of their work. As a result, auditee defensiveness in the face of questioning or perceived criticism is not uncommon. The lead auditor can head off potential confrontations by describing this dynamic at the opening meeting and counseling all involved in the audit to step back and take a deep breath if a situation gets tense. Appropriate humor also can help keep the audit in perspective.

Relations between auditor and auditee should be professional. Friendships can develop in an atmosphere of hard work and mutual respect, and personal discussions may arise during long drives between audit sites, but socialization outside working hours during the audit—even between long-time friends—should be avoided. Auditees also should remember that they are always "on." Informal moments can be used by the auditee to explain the organization's management in more detail or to contrast it graciously with that of other lands passed on the road, but too much "spinning" can be counterproductive.

The lead auditor assigns all specific auditor duties and tasks and is the team's final authority for operating decisions, final audit findings, the audit report, and recommendations regarding certification. Team members should make no independent contact with the auditee or any employee of the auditee unless specifically directed to do so by the lead auditor. The lead auditor must make it clear—this point cannot be overemphasized—that auditors cannot offer the client or auditee any advice on designing a sustainable forest management system or correcting nonconformances, whether before, during, or after the audit.

In addition to gathering evidence, auditors might be asked to review materials prior to an audit, to participate in a preliminary site visit, to serve as an assistant lead auditor with specific responsibilities in specific circumstances, to draft particular nonconformance reports, to make presentations in the closing meeting, to help formulate the certification recommendation, and to help draft the audit report. Depending on the audit firm, auditors with a high degree of experience might work alone in the field without supervision.

A technical expert might be asked to review materials prior to an audit and perhaps make a presentation at the closing meeting but should never work alone without supervision. A trainee likewise should never work alone as a trainee; an experienced auditor, however, might gather evidence alone but be supervised when practicing the role of lead auditor.

Good communication between the lead auditor and the team members is critical. Audit team assignments should be clear, within the capabilities of each auditor, and able to be implemented within a specified time frame. In return, auditors should express to the lead auditor any reservations about their ability to perform as expected—promptly, before difficulties arise.

Working within the scope of their assignments, auditors ask direct questions to appropriate auditee staff and must often be creative and persistent in eliciting responses. If staff members are not forthcoming and the team member is unable to get the information needed, he or she should seek the assistance of the lead auditor or assistant lead auditor. Team members should advise the lead auditor quickly about the possibility of a nonconformance—as soon as an office interview is over, say, or when a split audit team regroups at the end of the day.

Finally, auditors are hired for their expertise and skills, not for their personal opinions about forest management or other matters not pertinent to the conduct of the audit; such opinions are best kept to oneself in the presence of the auditee.

Preliminary document review

The audit firm will have gathered some information while preparing the bid for the audit, but the purpose of the preliminary document review is to find out more about the auditee to help in the preparation of the audit plan and save time during the preliminary onsite evaluation. The lead auditor needs to get a sense of the completeness of the auditee's forest management system and identify missing information. This review should begin as soon as possible after contracting for the audit. Important information that assists in audit planning includes the auditee's staff structure, the number and location of field offices that might be visited, the level and variety of field activities, and the accessibility of potential field sites.

The extent of the materials that can be reviewed remotely is limited by the availability of physical or electronic documents; the copying, mapping, and organizing capacity of the auditee; the availability of the auditee's staff for telephone interviews; and the confidentiality of the requested information. Normally this review is carried out solely by the lead auditor, but materials also might be reviewed by other audit team members if the documents can be made readily available, are particularly voluminous, or require specialized expertise to evaluate.

Audit plan

The content—or even the existence—of the audit plan varies by audit firm. However some certification schemes have minimum requirements as to the content of the audit plan. Ideally, the composition and roles of the audit team, the scope of the audit, its objectives and criteria, and a rough audit schedule are determined at the time of contracting and included soon thereafter in a draft audit plan. In any event, the essentials of a draft audit plan should be available for discussion at least by the onsite preliminary review. The audit plan should identify, at a minimum, the audit scope, objectives, and criteria, as well as the audit schedule and responsibilities.

Audit scope

Even before the final selection of the audit team, the audit firm and lead auditor determine the scope of the audit in cooperation with the auditee and state it in the audit plan. Dimensions of scope include geographic, organizational, and functional factors. An audit might cover a company's operations in North America, its operations in a state or region, or just those operations that serve a particular mill. The organization whose operations will be audited might be the entire company or just one division. And the audit might cover land management, procurement, secondary processing, or chain-of-custody for use of an on-product label. Thus in the written plan, the scope of the audit would appear as, for example,

- ABC's procurement systems in North America;
- ABC's land management systems in the XYZ Division; or
- ABC's chain of custody of FSC materials in secondary manufacturing in the northeastern United States.

Note that the scope of one certification might include the scope of more than one individual audit, as when various regions are audited separately but covered by one certificate. Moreover, individual certification schemes may place restrictions on the definition of audit scope for a public certification audit, particularly along functional lines, as in requiring that no public claim can be made on an audit to only part of the standard. For example, SFI does not allow an auditee to certify a mill's procurement operations without including its related land management in the scope. Audit firms should be alert to any unnatural delineation of audit scope, especially one that appears to be designed to avoid the auditee's problematic operations. For example, an auditee's attempt to exclude one forest district should be a red flag.

Contrary to a common assertion by auditees, contractors' activities must be considered to be within the scope of the audit. ISO 14001 (2004) uses the phrase "aspects of its activities, products and services…that it can control and those that it can influence" (4.3.1.a – see also 4.4.6. c) to help determine audit scope; that wording is a useful guide for all certification schemes in settling on the scope of an audit.

Group audits present a special case of scope, in that there must be clear criteria to distinguish members of a group from other landowners who might also have a relationship with the group manager but choose not to be involved in the group. How group members are distinguished from non-members should be described in the group's policies, and implementation of the policy should be documented for each member. The process by which group policies guide management for each member should also be clearly set forth.

Audit objectives

The audit objectives also should be stated in the audit plan. The statement of audit objectives often is underappreciated because *certification* is relatively well defined and the desired end of most publicly acknowledged audits. This particular objective should be stated as, for example, *an independent third-party verification of conformance to the PDQ Standard, for the purpose of certification.* There are, however, many other possible objectives for an audit:

- Verification of compliance with the provisions of law;
- Improvement through a consulting analysis of gaps in conformance and how to remedy them;
- An independent, third-party gap analysis without consulting but in preparation for a certification audit;
- An external, second-party verification of conformance to a buyer's, creditor's, or investor's environmental standards; and
- Surveillance monitoring or recertification of an existing certificate.

Note that the audit objective is related to the client–auditee's overall goals for the audit, and the audit manager and lead auditor should ensure that the relationships between these goals and the stated audit objectives are understood. They also should ensure that the audit objectives can be accomplished within the provisions of the certification scheme in question. There are three special considerations in defining the audit objective:

- The audit objective can become complex, as when two certification standards are involved, especially when, for instance, one is added anew during the surveillance audit of another.
- At least some of the audit objectives could apply to only part of a certification standard, even when that may not be allowed for a publicly recognized certification audit.
- ISO 14001 (2004, 4.2.c) requires compliance with applicable laws and regulations and "other requirements to which the organization subscribes," such as those of a forest certification scheme, even if the organization is not otherwise audited to that scheme.

Audit criteria

Closely related to the audit objective is a clear statement in the audit plan of the criteria that will be used to judge compliance or conformance:

- The version and/or date of the standard;
- The applicable type of the standard, such as small landowner, group certification, or green labeling;
- The applicable regional version (if any) of the standard;
- The applicable portion of the standard (if not the entire standard);
- The levels of the standard that apply or are considered mandatory,
- Any implications for the binding nature of the auditee's own policies, procedures, and plans (which are generally that they also must be followed), and
- Any other voluntary indicators or other audit criteria selected by the auditee.

In addition, the audit plan should state the manner in which any nonconformances with the criteria will affect certification or other audit objectives. The choice of criteria are essentially up to the client (who selects the standard, may add criteria, and may propose that some will not apply) and accepted by the auditee, but the statement of audit criteria and their use also must be approved by the lead auditor as consistent with the statement of audit objectives and with the certification standards and scheme procedures. Some certification schemes have interpretations that are published separately from their standards and procedural guides to provide additional direction on the appropriate scope and conduct of audits.

Forestry standards are often referred to as *sustainable forest management* standards or as measures of *good management*. Such judgments are subjective, however, and are meaningless without a clear statement of the certification scheme's criteria. In the end, forest certification is nothing more or less than an external verification that the auditee's activities within the scope of the audit conform to the stated audit criteria.

Preliminary audit

A preliminary audit or onsite visit is typically conducted by the lead auditor alone or with perhaps one other auditor. It provides an opportunity to review materials not available for remote review, interview auditee staff members in person, and engage in more complex (and often more confidential) discussions about the auditee's operations. The draft audit plan can be reviewed and refined and apparent deficiencies can be discussed, clarified, and often rectified. Such a review typically takes about two days but can run a full week, especially if field sites are visited. For very simple or return audits, the preliminary meeting might be completely forgone, especially if the remote document review is quite complete.

For efficiency's sake, the onsite preliminary review should occur after the lead auditor has had ample time to review the auditee's documents but far enough in advance of the field audit that the auditee will have time to correct any deficiencies. Timing can be tricky, however, because of the need to coordinate everyone's schedules for the audit in an atmosphere of uncertainty about the readiness of the auditee. It is important, therefore, to realistically assess readiness early in the audit process. This can be done most effectively with a preliminary gap analysis, conducted either by the audit firm or by another firm on a consulting basis, before the field audit is even scheduled.

In reality, however, some auditees choose not to incur this additional expense or delay. Contract provisions often require the auditee to pay a penalty for audit rescheduling to at least partly compensate the audit firm for costs incurred and a lost opportunity for revenue. Such pressures for profitability from both the audit firm and the auditee typically lead to a strong preference for a "tentatively" scheduled audit, despite uncertainty—and even serious doubt—about the readiness of the auditee's documents, systems, training, and performance.

Although the audit firm has the option of declaring the auditee unready for the audit and the auditee can choose to delay the audit, the lead auditor should at least be clear about perceived nonconformances and challenge the auditee's staff to realistically assess their ability to remedy the existing deficiencies in time for the field audit. The auditee can then weigh the costs and benefits of proceeding with an audit whose outcome is questionable. Auditees may decide to go ahead with an audit and plan to correct nonconformances afterward, sometimes to motivate the organization to make needed changes or just to find out where they stand. (ISO and CSA Z809 audits, however, cannot be scheduled until the auditee has had its management system in place for a year or for one cycle of internal audit and management review.)

Field audit

The field audit is actually a complete system audit of documents (including those in field offices), staff, and field operations and results. Most of audit team's time is spent in the field, however, where the operation of the auditee's system is explained in application and the actual results are shown. Little general can be said of field audits because the situations they cover are so highly variable, but they typically range from 2 days to 2 weeks.

Any factor that creates more variation in performance (e.g., ecological variation, complex management systems, or highly variable weather) requires additional sampling to account for that variation. The factor most affecting the length of an audit is the number of field offices that must be covered, especially if the organization is not centrally standardized or controlled. Differing field office operations can be a significant source of variation in performance which requires more intensive sampling.

Most audit firms favor continuous reporting of their preliminary findings during the course of an audit, thereby giving the auditee an opportunity to present additional evidence, interpret data differently, and challenge audit results. This approach is facilitated greatly by a routine of daily opening and closing meetings—preferably involving the entire audit party in one location—to discuss findings, confirm audit plan details, and adjust plans as needed to challenge or verify tentative conclusions. Auditors must remain in contact and exchange information on their findings as often as is practicable.

Field auditing for forest certification is particularly complex because the auditor cannot be present for the whole chain of events—often several years long—that produced the results being observed at one point in time. As a result, the audit team needs both documentation and personal explanation of plans and accomplishments to understand and evaluate what is actually observed on the ground. This process is even more challenging when the team is split for efficiency's sake: Not all members—and hence not all specialized expertises—will be present for all field site visits, even though any element of the standard could potentially apply anywhere. To balance efficiency with broad coverage, therefore, audit team members must brief each other on particular things to look for pertaining to each element of the standard. Finally, the entire audit team can participate in the first few field site visits to hear the auditee's explanation of the field management process in relation to elements of the standard.

The site visitation plan is determined by the lead auditor or audit team (or both) and based on the entire range of conditions and practices on the forest in question over the time period of the audit. The plan can be changed during the course of the audit in response to apparent observed problems. Field sites should be selected with the following objectives in mind:

- To cover adequately all audit issues, major forest areas, and administrative jurisdictions;
- To give special attention to sensitive issues;
- To ensure that team members visit sites with issues related to their areas of expertise; and
- To minimize travel time between sites.

Because time is of the essence in field audits, logistics should be designed for convenience. Auditors spend long evenings documenting, compiling, and discussing findings, so any extra time auditors must spend in traveling to luxurious accommodations, eating fancy meals, or just socializing may not be appreciated.

Collection of evidence

The essence of auditing is collecting objective evidence and evaluating it against an objective standard. "Objective" evidence is any information, quantitative or qualitative, relevant to the standard that could be verified reliably by anyone with equivalent access to that information and adequate expertise to evaluate it. The result of comparing objective evidence against an objective standard should be a replicable audit.

In any kind of audit, expertise in the subject matter is necessary so that the auditor can identify or know where to look for evidence. This is especially true for a forest certification audit, given the complexity and variability of forests: Local expertise is critical. In addition, the standard may include elements that are not clear or for which it is difficult to obtain objective evidence and for which an unusually high degree of judgment is needed to identify, interpret, and evaluate evidence. This makes the forest certification auditor's job much more difficult and requires balancing the intent of the standard with the literal—and sometimes vague—language of the standard.

Objective evidence comes from three types of sources:
- The auditee's documents and records;
- Interviews with the auditee's staff; and
- Direct observation of conditions in the forest, mill, or other sites.

These information sources roughly correspond to, in order, the intent of the organization as expressed in policies, procedures, and plans; implementation of that intent; and results or performance of that implementation. But note that documents also can demonstrate the implementation of policies, as can results not immediately observable by an auditor. Interviews can demonstrate, confirm, or explain the existence of a policy, procedure, plan, or result. And a direct observation makes sense only in the context of the specific implementation of policies, procedures, and plans.

To aid in objective decision making and to be able to explain their audit results, auditors must systematically gather and document the evidence they use to make decisions about conformance. Different auditors and audit firms use a wide variety of techniques and tools to document evidence, including checklists, matrices, interview forms, copies of records, annotated copies of standards, and simple pads of paper. An evidence list based on the standard is often provided by the auditee and can become an evidence checklist and, especially if in digital form, a repository for notes on audit evidence.

Findings should be documented in a format that is amenable to summarization in terms of site locations and the coverage of issues. Auditors find it easier to identify documentary evidence if it is labeled with references to particular elements of the standard (usually using the standard's numbering structure) and organized in a binder or set of binders, preferably one for each audit team member. Even then, however, one piece of evidence or different pages from the same document might relate to several elements of the standard, requiring either duplicating the same document or cross-referencing its occurrence at another location in the evidence binder. Similarly for records that have multiple pieces (such as stand prescription forms or harvest inspection checklists): Usually only one or a few examples in the binder can be backed up by a reference to particular file locations where the remaining copies of that class of records can be found. In part to overcome these problems, documentary evidence is often now provided on a compact disk or via access to an intranet website, often with hyperlinks leading from the evidence list written into the text of the standard and vice versa.

The auditee may choose to structure policies to the audit standard or to some other outline or format (which makes particular sense when the auditee seeks certification to more than one standard). In either case, each piece of evidence should be referenced to the element of the standard to which it relates. Especially in the case of more than one standard, electronic hyperlinks to digital documents make complex cross-references simpler, but this may make it harder to view multiple related documents at one time. However documentary evidence is to be presented, it makes sense to work with the auditee early on to arrive at a format that fits both the auditor's and the auditee's needs. Finally, arrangements also should be made early in the process if the auditor would like to retain documentary evidence in either hard copy or digital form for the auditor's files.

Interviews are an important component of evidence because they provide links between the auditee's policies and the actual performance outcomes. They reveal staff knowledge and skills and explain specific decisions. Interviews with regulatory agencies or other stakeholders also can provide important independent information about the auditee's performance. Information acquired in interviews should be related to specific elements of the standard in question and documented in some consistent manner.

Findings and conclusions

Once objective evidence has been collected and evaluated against the audit criteria, a judgment must be made about whether the auditee's management conforms to the standard. This determination must be made based on the evidence, the standard, and an understanding of the subject matter of the audit—not on personal or professional preferences for particular management systems or outcomes.

Conformance means, literally, "with shape"; it denotes agreement between the form of a standard and the form of the auditee's system. A finding of conformance means that the auditee's management is in the shape required by the standard. It may help to know that dog shows are called "conformance events," and the dogs are examined by judges to see whether their appearance, muscles, bones, and coat conform to the breed standard. Similarly, in a forest certification audit, objective evidence is examined and compared with an objective standard to determine whether the auditee's forest management conforms to the standard.

The idea of conformance must be distinguished from that of "compliance", which should be used only to apply to legal compliance. Although all standards require auditees to comply with applicable laws, forest certification audits are not legal compliance audits, and this distinction must be maintained to avoid the assumption of legal liability by the audit firm for the legal compliance of the auditee. Instead, forest certification audits include only a review of the auditee's systems for ensuring compliance, not an assurance by the auditor whether compliance, per se, has occurred.

There are three common degrees of conformance:

- *Conformance*, a finding that the entire management system—policies, schemes, procedures, and field implementation—is in the shape required;
- *Minor nonconformance*, a finding of a slight or isolated deviation from conformance that does not affect the overall viability of the management system; and
- *Major nonconformance*, a finding of systematic failure in a system.

A major nonconformance can originate in many ways, including the lack of a required policy, a missing or nonfunctional procedure, or an accumulation of several minor conformances that together indicate that a procedure is not producing reliable results. Although the number 3 has a certain intuitive appeal, no set number—or for that matter, proportion—of individual minor nonconformances can constitute a clear threshold of major nonconformance. Each case requires a judgment by the auditor about the systemic nature of the nonconformance; an auditor may wish to increase the sample size for the category in question—usually by decreasing the planned sample size of other criteria for which conformance has been demonstrated.

Virtually by definition, *certification* cannot be conferred in the event of a major nonconformance. In some traditional FSC terminology, major nonconformances may be called *preconditions*, meaning that a remedy must be implemented before certification can be granted. Minor nonconformances, or "conditions" in traditional FSC terminology, mean that remedial plans must be implemented by a certain time after certification. Some certification schemes use numerical scores for each standard element, which are then related to thresholds for major and minor nonconformances.

Some auditing firms also use terms like "observation" or "opportunity for improvement" or even "recommendation" to refer to a potential for a future nonconformance that does not require a current formal remedy. As with the finding of a nonconformance, particular care must be taken in the use of such categories of findings to avoid any suggestion of consulting on the nature of the remedy (as opposed to merely pointing out the existence of a nonconformance) or inappropriately using the lesser category of finding to please the auditee. Finally, it should be noted that all audit results—even of conformance—are "findings", even though some auditors are in the habit of using the term to mean only findings of nonconformance.

Whatever the terminology, all audit findings are, in a sense, tentative. A finding of conformance, for example, is based on a sample that has as yet yielded no deviation from the standard, but one more observation could yield the opposite result. A finding of nonconformance, on the other hand, may be based on the lack of a document that has not yet been found, a single staff member who does not know a certain policy, an assumption about a prior resource condition, or a misinterpretation of a fact: An additional piece of evidence found by the auditor or presented by the auditee could reverse the finding. For this reason, conclusions about conformance depend on the decision that enough observations have been made and that tentative findings have been adequately reviewed by the auditee.

As any tentative findings of nonconformances accumulate, the lead auditor summarizes them and may comment on their potential seriousness as major or minor nonconformances and possibly on their cumulative impact on the overall audit conclusion. It is most convenient to do this at the daily opening or closing meetings, when the entire audit party is together and after the audit team has had a chance to confer and concur on their findings. The auditee staff can then respond with additional evidence or alternative interpretations, usually after taking the time to confer among themselves.

Before finalizing an audit finding as a conclusion about conformance, a decision about the adequacy of sampling and auditee review must be made. This process is most effective if not delayed: Tentative findings and interpretations are shared with the auditee as soon as they begin to be formed, tested with new observations, and elaborated in discussions with the auditee. When the auditor is convinced that additional information is unlikely to reverse the tentative finding, the tentative findings are formalized as an audit conclusion. This is essentially a sort of statistical statement that balances the consequences of a possibly wrong finding with the additional cost of making that finding more certain. The tradeoff is made more easily if the auditee has progressively more formal opportunities to challenge a finding of nonconformance, with each step involving more clarity and greater understanding about the logic of the auditor's conclusions.

Reporting

To facilitate communication between auditor and auditee, audit findings are typically presented to the auditee at the time of detection, in summary each morning and evening, and at the closing meeting, before finally recording them in written draft and final reports. In addition, audit conclusions may be presented to the public, generally in summary form, as may be required by the certification scheme.

After conferring in evening meetings, audit teams often write up findings of nonconformance on standard report forms. These forms are commonly if misleadingly called corrective action requests—CARs—because they initiate a remedy process. Their purpose is to notify the auditee of nonconformances and, especially, to clarify the logic justifying the findings. "Corrective action request," unfortunately, suggests that the auditor is requesting a specific remedy; remediation of a nonconformance, however, must be undertaken at the initiative of the auditee. It is important for auditors to avoid the appearance of suggesting a remedy, especially as auditees are naturally inclined to want free consulting and to arrive quickly at a remedy that the auditors will accept.

Whatever they are called, written findings of nonconformance provide a basis for a clear discussion of audit issues. Nonconformance forms might include a signature line on which the auditee recognizes or accepts the finding of nonconformance and could also have space for the auditee to propose a remedy or record a remedy approved by the auditor and for signatures formalizing this agreement. If a finding is challenged by the auditee, these forms may be redrafted to more adequately express the nature of the nonconformance. A summary form recording all nonconformances also is useful for keeping track of them.

Whether or not nonconformances are written up before the audit's final closing meeting, they are typically presented there in final form, but only after an oral presentation of overall audit findings, both positive and negative. The auditee's proposed remedies also might be discussed there, along with a timetable for completing remedy proposals, and their approval, implementation, and verification.

After the audit (sometimes right after the closing meeting), the lead auditor or the audit team draws up an audit report. Depending on the audit firm, individual auditors might write portions of the report or review the entire report. FSC requires audit reports to be reviewed by a group of external peer reviewers, and all major schemes require audit firms to have their own internal review processes as well. Sometimes reports are then reviewed by the auditees for matters of factual accuracy before being finalized by the lead auditor.

Minimum audit report contents are dictated by each certification scheme but typically include a description of the auditee; a restatement of the audit scope, objectives, criteria, and team; and the methods, schedule, and findings. Reports also may include nonconformance remedies and their verification, or these items might be an addendum to an earlier preliminary report.

The audit report alone, however, is not the sole determinant of the final certification decision. Typically, the report makes a recommendation about certification to the audit firm's audit manager, who then reviews the report (possibly with the assistance of external peer reviewers or an internal review team, or the input of the auditee) as the basis for the final certification decision. The audit manager then delivers the final audit report, accompanied by the conformance certificate if the auditee "passes," to the client.

Finally, depending on the procedures of the certification scheme and the audit firm, a public report based on the audit report is drafted, usually with more auditee involvement but in less detail. Public reports are then made available on request or posted on the website of the auditee or certification scheme.

Nonconformance remedies

Whether in the course of an audit, in clearing a major nonconformance before certification, or in following up on a minor nonconformance in a later surveillance audit, an auditor must determine the adequacy of a remedy proposed and implemented by an auditee in response to the finding of a nonconformance. This determination inevitably involves an inference about future behavior based on a somewhat limited sample of very recent behavior. This is not unlike auditing the performance of any new forest management system except that, by definition, the auditor is looking for convincing evidence of a change in behavior and performance. At the same time, a delay in the determination of a successful remedy could result in a suspension or delay of certification. Because of a need to balance these two factors, remedy proposals and their implementation must be examined carefully and, often, more evidence must be gathered on a timely basis to arrive at a reliable conclusion.

Appeals

Not all interested parties will necessarily be satisfied with an audit decision. The auditee—or perhaps an auditee's logging contractor—might believe that the audit firm was too strict, or a stakeholder or critic might consider the audit firm too lax. Appeals processes provide an opportunity for potentially aggrieved parties to seek remedies and an incentive for auditors to conform to the audit firm's expectations.

Most certification schemes require that its audit firms establish formal appeals procedures to deal with objections from the audit client, although other informal procedures may also be used. In addition, some schemes provide their own procedures to deal with others who question the validity of the audit. Finally, each entity also might have procedures to deal with the other category of claim, the certification scheme could provide a second level of appeal after that of the audit firm, or aggrieved parties might present their objections to the accreditation body of the audit firm or the certification body of the auditor (if separate from the organization of the forest certification scheme).

Audit appeals must be clearly based on the grounds for appeal as delineated by the appeals procedures of the audit firm or certification scheme. Such grounds might include overlooked evidence, auditee dishonesty, audit procedure irregularities, or an auditor's lack of competence or conflict of interest. Appeals claims are then evaluated according to specific appeals procedures, which might include remanding the issue to the lead auditor or audit team for reconsideration or appointing another auditor to investigate the problem. If the appeal is found to have merit, remedies could include granting a certification that had initially been denied or revoking a certificate already granted. Presumably, contractual relations between the audit firm and client or specific interests of third parties could provide a basis for legal action regarding the outcome of an appeal.

4 Forest Certification Case Studies

The concepts in this book are intended to provide readers with an overview of forest certification auditing, including the history, terminology, philosophy, and techniques of auditing, for the major forest certification auditing standards in use in 2007. As in most technical subjects, learning often is best facilitated by putting readers in situations that require the application of the content learned. The best learning environment for understanding forest certification auditing is through direct observation of and participation in audits, but a good substitute is the use of case studies.

The 15 case studies provided, using the SFI, FSC, and American Tree Farm System standards, illustrate a range of situations that a forest certification auditor might encounter. Although they cannot be comprehensive, these situations should give prospective auditors or auditees valuable information about what will be encountered on a forest certification audit. Each case is based on a real audit, but company and forest names and other particulars have been altered to ensure that no confidential information is revealed.

Once readers have completed reading a particular case, they are directed to search the appropriate forest certification standard for the sections that apply to the situation in the case. Current versions of the standards are available at the following websites: SFI, www.sfiprogram.org/sfistandard.cfm; FSC, www.fscus.org/documents/; and Tree Farm, www.treefarmsystem.org/cms/pages/26_19.html. Readers should have a clear understanding of the relevant portions of the standard before answering the questions posed. Brief answers are given starting on page 47.

SFI Case Studies

Case Study 1.
Precertification: Preliminary Meeting

You are a member of the management team at the Portnoy Pulp and Paper Company in Georgia and are taking part in the preliminary meeting with the auditor your company has hired to carry out a third-party audit of your company's conformance to the SFI Standard. The company CEO has decided to seek SFI certification and has hired the auditing firm but at this time does not plan to use the SFI on-product label. With you at the meeting are the lead auditor for the auditing firm, another auditor hired by that firm, and Portnoy's CEO. The CEO confirms with the lead auditor that the audit is going to cover only its land management operations, because he won't be labeling his paper.

You head up all the forestry operations on the company's 114,000-acre forestland ownership as well as wood purchasing for the pulp mill. Portnoy purchases 80% of its annual 2 million tons of wood from nonfee lands. During introductions, you learn that the lead auditor is a licensed forester, SAF member, and a RABQSA-certified environmental management systems lead auditor. The other auditor is an accountant from a local CPA's office. At the meeting, the lead auditor hands out his résumé and the payment schedule for the audit. He says that, depending on what they find during the precertification investigations, he may or may not add a third auditor to the team for the field audit. After initial discussions about what the two-person team will be doing on-site, the two auditors ask to see company documents related to the SFI Standard, and you begin to provide them.

Questions
1. Is there anything else that should have been discussed at the preliminary meeting? Is there anything else the lead auditor should have provided in the way of documents?
2. Is the audit team adequate?
3. Is the audit scope clear?
4. Is the audit scope consistent with the standard?

Case Study 2.
Field Audit: Sustainable Harvest Levels

You are part of a four-person audit team assigned to an SFI third-party audit of Polbear Forest Products, a Maine-based sawmill and forest landowner company. Polbear owns 15,500 acres in central Maine and also procures wood for its mill; about 30% of its wood needs come from Polbear-owned forestland. During the precertification investigations and document review, you learn that Polbear doesn't have very good timber inventory data for its fee-owned lands. A cruise was conducted in 1982 as part of a sale package (Polbear owned no land prior to that), but since then, no inventory work has been done. The company's growth-and-yield calculations are very simple, done by the company forester on a 1-page Excel spreadsheet. She took the 1982 inventory data, subtracted the harvests over time and a percentage for mortality (based on data from Maine's annual Forest Inventory and Assessment (FIA)), and then made a determination of growth using the most recent FIA data for the county. Based on this calculation, Polbear has been harvesting at growth for the past 10 years, after having reduced the stocking in the first 10 years of ownership from 22 cords per acre to 17 cords. In the field audit, you visit 43 sites over 3 days' time, and based on the stocking you've seen in the woods, you are concerned about whether the harvest levels are sustainable.

Questions
1. Is Polbear conforming to the section of the SFI Standard on sustainable harvest levels? What are those references in the standard?
2. Are the data adequate relative to what the standard requires?
3. Could you, as an auditor, have done anything else in gathering evidence to better determine whether conformance was being demonstrated?

Case Study 3.
Precertification Investigations: Document Review

You are a member of a team of four people performing a third-party audit on the Woodgood Company, owner of 230,000 acres of forestland in the Lake States. Woodgood owns no manufacturing facilities and doesn't procure any wood. Company foresters manage all of Woodgood's land and market cut timber to various markets. All harvesting is done through contract loggers on a per-volume rate. No stumpage is sold.

During a precertification document review at the company headquarters in Wisconsin, the lead auditor has asked you to investigate Woodgood's conformance with the reforestation portion of the SFI Standard. You have been provided a stack of documents on reforestation by Woodgood's forest manager. Your review of these documents reveals the following.

The company's reforestation policy reads, "Woodgood ensures that areas where final harvests occur are reforested through either natural regeneration or planting within 5 years from the harvest date." Company foresters rely mostly on natural regeneration but also plant some sites, usually with species that are faster growing than native species.

There is no written procedure for judging regeneration success. According to the manager, at year 4, the forester in charge of the harvest site visits the area, walks into the cut, and looks around. If he sees what he believes to be 1,000 or more seedlings growing, he knows regeneration is adequate. If not, he runs 1/1,000-acre plots on a grid to determine the number of seedlings present. This procedure has been the company way since before the manager can remember.

Questions
1. Does Woodgood's reforestation policy conform to the SFI Standard?
2. Is the method for determining regeneration adequate?
3. Is the company conforming to the standard's requirement on "exotics"?
4. Is there anything else in the way of documentation that you should investigate relative to SFI reforestation requirements?

Case Study 4.
Precertification Investigations: Determination of Readiness

The Laguna State Department of Forestry has contracted with Pearmont Registrars to undertake an SFI certification audit to determine conformance to the SFI Standard for its 900,000 acres of state land. The state is a brand-new licensee under SFI. After initial investigations by Pearmont, it is clear that the department has not adopted the SFI Standard as its forest management system framework. Furthermore, document review suggests that many required written policies do not exist, the Laguna SFI Implementation Committee hasn't met for more than a year, and the loggers who harvest on state lands—the same companies that have done so for almost 20 years—are not trained in best management practices or through the state's logger training course, which is offered by the University of Laguna. Despite the lead auditor's determination that the department is not ready to undergo the full third-party audit, the Laguna state forester decides he wants to proceed with the audit anyway as a way to firm up a "gap analysis" of the department's efforts relative to the SFI Standard.

Questions
1. Is it appropriate for Pearmont to complete the third-party audit even though the lead auditor has determined that the Department of Forestry is not ready for it?
2. Will the results of the audit really be a third-party audit if, indeed, the Laguna state forester will be using the results as a gap analysis?
3. If the audit by Pearmont results in several major nonconformances, does that mean that the state lands can't be certified, as suggested by Pearmont's readiness determination?

Case Study 5.
Field Audit: Best Management Practices for Water Quality

You are a district forester for a 120,000-acre ownership in Tennessee. Your company prides itself on its best management practices (BMP) training for staff and its loggers and road crews. Company foresters regularly review BMP compliance through harvest and road construction close-out reporting. The company's incentive program for contractors is well-known, and those contractors with the highest scores on close-out inspections receive 5 percent bonuses.

The company is being third-party audited to the SFI Standard. The audit has gone well so far: The audit team found nothing untoward in two other districts. The auditors are now in the field with you, reviewing your district management. They decide at the end of a long day to see the farthest eastern section of your district, where a half-mile section of spur-road was built last spring. Neither you nor any of your foresters have been there since a harvest took place in July and was closed out. Two weeks ago this part of the state received 6 inches of rain in 2 days.

You head around the first turn in the spur road, where a hill begins and the road crosses two streams. Both culverts at the stream crossings are blown out, and parts of the road have been washed out. Significant siltation is evident in the stream bed, and water is flowing across the road through the ditches where the culverts used to be. The auditors ask you to stop and begin their walk to check out the road and culvert problem. After a 5-minute inspection and a discussion during which you explain the circumstances, the auditors suggest that they will be issuing a major nonconformance.

Back at your office, you make copies of all your records on monitoring for BMP compliance on roads and logging operations and give them to the auditors the next morning.

Questions
1. Was a major nonconformance warranted?
2. Is there something you could have done to anticipate the situation?
3. Were the auditors being unreasonable?

Case Study 6.
Opening Meeting

The opening meeting for an SFI third-party audit begins at the Chenawgwa Forest Products Company in California. The company has a sawmill and secondary manufacturing plant as well as 73,000 acres of timberland. It wants to use the SFI label on its finished products. Three auditors are present from the auditing company hired by Chenawgwa. After introductions, the lead auditor and the Chenawgwa contact, the vice-president for forestry, discuss the scope of the audit. The vice-president suggests that the audit focus on just the mill procurement operations and the 40,000 acres near the mill site, but if the field manager returns from vacation in time, they could add his 33,000 acres. Because Chenawgwa had a mill fire the week the lead auditor had planned to come for a precertification meeting and document review, she has not done any document review prior to the opening meeting. She wants her team to spend a day reviewing documents before getting out in the field.

When the group begins to review documents, the auditors are denied many of the files and documents they request because access hasn't been cleared by the field manager—the forester on vacation. The audit team gathers what it can before heading into the field on day 2 of the 4-day audit. The field forester in charge of the 40,000-acre district has been on the job only 6 months and doesn't know a lot. He suggests they wait for the field manager to return because he "has all the answers."

Questions
1. Is the agreement between the lead auditor and the company vice-president adequate?
2. Is the field manager's absence during the audit a problem?
3. Can you identify any flaws in what was covered in this opening meeting?

Case Study 7.
Critically Imperiled and Imperiled Species

You are auditing a 24,000-acre property in Oregon owned by a privately held company. You are the wildlife biologist–ecologist on the auditing team, and the lead auditor has asked you to focus on the special sites objectives. You have reviewed the information provided by the company about its policies, programs, and procedures on wildlife habitat, threatened and endangered species, and critically imperiled and imperiled species.

The documents say that the company will identify and protect all threatened and endangered species on state lists in addition to the federally listed species, and it appears to be doing just that. Regarding G1 and G2 species, the company forester tells you that there are no known sites on the property, so the company doesn't need to address anything except threatened and endangered species. When you say that Objective 4, Performance Measure 4.1 suggests otherwise, he tells you that that only affects properties outside the United States and Canada.

Questions
1. Has the company provided adequate evidence that it is addressing the G1 and G2 species requirements?
2. Is the forester correct in his assumptions about those requirements?
3. What else does the standard require for G1 and G2 species conservation?

Tree Farm Case Studies

Case Study 8.
Management Plans

A consulting forester from Wisconsin is seeking certification of her client base under the Tree Farm group certification process. The client base for this forester is diverse: Some landowners own more than 5,000 acres and others as little as 15 acres; some clients have been with her since she started her practice 21 years ago and others are new clients in the past year.

Most of the clients in this consulting forester's business have formal written management plans developed by the consulting forester, but several long-standing clients do not—although the forester says there is agreement on the management direction for the property. A few of the new clients also do not have written management plans, and she says that she will develop plans for those clients. She has an overall business plan and forest management philosophy that addresses issues normally found in management plans. She believes that her contract with each client amounts to the equivalent of a formal management plan because one clause commits the client to the management philosophy and approach contained in the business plan.

Questions
1. Does this consulting business have a group that is eligible for Tree Farm group certification?
2. Has the consulting forester demonstrated conformance regarding management plans?
3. What should the management plans contain?

Case Study 9.
Type of Group and Eligibility

Potmar Corporation procures sawlogs over a wide area in western Kentucky, and its procurement process is certified to the SFI Standard. To increase its appeal to landowners, Potmar offers forest management plans to landowners for a small fee, which is waived in exchange for a right of first refusal on timber sales. Potmar would like to have this program certified as a Tree Farm group to make the wood more valuable for its SFI program: The wood will be considered more directly certified than if it had just come through the company's procurement system, and several local markets are giving preference to Tree Farm wood.

Questions
1. Is this group eligible for Tree Farm group certification?
2. Does the company's right of first refusal on landowners' timber affect the potential to have this group certified under the Tree Farm scheme?
3. What other observations do you have about this situation?

Case Study 10.
Special Sites

You are on a certification audit to determine whether a group maintained by the Yahoo Forest Landowners Consortium conforms to the Tree Farm group certification standards. As the lead auditor of the audit team, you find yourself on the land of one of the 327 members of the group. You notice that the landowner has built a new pole barn and shed near the center of his 210-acre property to house a farm tractor and other equipment used on the property for management purposes.

The structures overlook a small river and also are near the road that runs through the property—the landowner's stated reason for choosing this site. Also, he jokes, the shed's back porch provides the perfect spot to hunt deer in the fall, with clear view of a major wildlife trail on the other side of the river.

You had read in the special sites section of the management plan that Native American artifacts and living sites had been discovered and documented along this side of the river. The landowner says he's not sure whether the structures are located in the area of the archaeological sites, but he didn't find any arrowheads in his post holes. According to the management plan, "every effort will be made to ensure that these archaeological sites will remain undisturbed," to which the landowner replies that there are other, untouched sites.

Questions
1. Is this situation covered by the Tree Farm standard?
2. Is this landowner conforming to that portion of the standard?
3. If not, what is the nonconformance? What might be an acceptable remedy to the nonconformance?

Case Study 11.

Reforestation

A landowner assistance program developed by a sawmill company in a midwestern state is seeking certification of its clients to the Tree Farm group certification standard. This state has no regulations regarding reforestation, although there are some requirements if a landowner is enrolled in the state's current-use taxation program. Under this program, property taxes are reduced from the highest-and-best-use taxation that would otherwise apply in return for the landowner's not developing the land.

A particular landowner's management plan indicates that an objective of his Tree Farm management is to ensure "adequate regeneration of all forest land harvested within 5 years with species that have economic value in the region." As the auditor of this 450-acre property, you notice that two 10-acre plots that were clearcut almost 7 years ago have very little regeneration. A native grass species has formed a solid mat on the ground, and it is clear that natural regeneration is going to be a challenge. When you ask the landowner whether these areas have "satisfactory stocking levels" as the Tree Farm Standard requires, he admits that they probably do not but says that he needs more openings anyway. When you ask whether this is consistent with the management plan, he replies that he is entitled to change his objectives. He adds that another forest owner friend had this grass problem, and it cost a lot to do proper site preparation to ensure adequate regeneration.

Questions
1. Can a landowner change his or her management objectives for the land?
2. Does this situation warrant a finding of nonconformance to the standard?
3. Is cost a legitimate reason to justify the landowner's failure to ensure adequate regeneration?

FSC Case Studies

Case Study 12.
Sustainable Harvest Levels

A family-owned company in the South is seeking certification under FSC standards. The ownership is about 500,000 acres with 300,000 in natural pine and pine plantations and the remainder in bottomland hardwoods. One question raised in the certification assessment has to do with forest inventory methods and the determination of allowable annual cut.

You and other members of the assessment team have observed that the company is probably achieving its commitment to sustainability of annual harvest with the present level of harvest. However, you are not convinced that this could definitively be established with the present inventory data and method for assessing growth-and-yield. The technique used for pine is best described as a stand projection method. Since the last full inventory of pine stands was conducted about 20 years ago, individual stand volume has been updated and projected forward, or estimated in the case of new stands, by using individual tree growth data from 440 trees stratified by site and age and a few 1/3-acre growth plots. Hardwood growth-and-yield has been calculated in a similar way. Stands have been inventoried only when observations by experienced foresters indicate that stand volume may be different than that projected by the model, but a full inventory of the pine and hardwood resource is scheduled to begin within the next few months.

Questions
1. Is the method used to determine growth-and-yield adequate?
2. Can allowable harvest levels be calculated sufficiently from the data available?
3. Is a condition (minor nonconformance) or precondition (major nonconformance) warranted?

Case Study 13.
Pine Plantations

State forestlands managed by a state forestry department in the South are the subject of an FSC certification assessment. The agency manages approximately 150,000 acres of natural stands of upland hardwoods, natural stands of pine, and pine plantations. The management of the pine stands, particularly the pine plantations, becomes an issue during the assessment.

Silvicultural methods used for the pine plantations, including site preparation and harvesting methods, are found to be satisfactory. The regeneration cuts (clearcuts) are often larger than 100 acres and sometimes as big as 200 acres. Although that is not desirable, size of the regeneration cuts alone would not prevent FSC certification or necessarily call for a condition (minor nonconformance). The agency meets the minimal state best management practices requirements for streamside management zones, but there is no retention in any of the large regeneration areas, and the agency's management plan does not address retention in regeneration cuts larger than 40 acres.

Questions
1. What portion of the FSC standard is relevant?
2. Is the size of the clearcuts a problem?
3. Is a condition (minor nonconformance) or precondition (major nonconformance) warranted here?

Case Study 14.
Management Plan

The current land base of a privately owned forest in the Midwest is approximately 160,000 acres, and the forest is oak–hickory upland hardwoods. The ownership has no manufacturing facilities and employs a staff of six. The stated mission of the ownership is to demonstrate how privately owned forestland can be successfully managed for the enhancement of timber, wildlife, water, recreation, aesthetics, and other resources using sustainable forest management practices through the application of uneven-aged forest management techniques. The land is managed to preserve the beauty of the forest while maintaining a continuing level of output from renewable resources. Managers of the forest are strongly dedicated to this mission and have long tenure on the forest—some for more than 25 years.

In the audit, you learn that, despite the size of the holding, no detailed management plan exists. An abbreviated forest management plan has been prepared, but it is rudimentary and does not contain many elements normally considered necessary for management of a forest of this size. Nevertheless, detailed planning has been done for most activities, the operation generally runs smoothly, and planned activities are accomplished in a timely manner. You have no concern about the long-term objectives or the ability of the staff to achieve them. However, their success largely is due to their accumulated knowledge, long tenure, and experience, and to excellent oral communication between staff members and between the staff and the landowner.

The abbreviated plan and other supporting documents do provide information on short-term and long-term goals and objectives; silvicultural systems used; rationale for the rate of annual harvest; and except for information on landscape-level considerations, a good description of the resource to be managed.

Questions
1. Is an abbreviated plan sufficient, given the staff's demonstrated expertise?
2. If you believe the plan is not sufficient, what parts are missing or inadequate?
3. What other sections of the FSC Standard are affected by this case?

Case Study 15.
Local Contractor Issues

A large forest landowner in the Northeast United States is undergoing an initial FSC certification audit. The company contracts regularly with two local firms for timber stand improvement work—largely brushsaw precommercial work to thin spruce and fir stands in the sapling size class. It has had a relationship with these companies for more than 8 years. Every fall, the company meets with the contractors to agree on the work volume, location, and price for next year's field season of timber stand improvement. A contract has resulted each year from those meetings and negotiations.

As the social expert on the FSC audit team, you learn from interviews that this past fall, the company decided not to hire either of the local firms, even though they had 800 acres to complete in the coming season, according to their annual work plan. Through a broker, the company has decided this year to hire a group of workers from Mexico. The president of one of the local firms says that the company executives weren't even willing to sit down to discuss whether he could renegotiate the contract or do anything to win the work for the coming season. Because this work was a major portion of the firm's warm-weather business, the loss of the contract is crippling.

Questions
1. Is this situation covered under the FSC Standard for the US Northeast?
2. What should the company have done, at a minimum, to meet the spirit and intent of the standard?
3. Is hiring foreign workers a problem in and of itself?
4. Does this situation warrant a condition (minor nonconformance)?

Some answers to forest certification case study questions
Case Study 1
1. Yes, there are two important items that should have been discussed at the opening meeting: first, the audit plan, an essential component of any good audit; and second, the indicators to be used in the audit. A discussion should also have ensued about the audit team and members' qualifications for conducting the audit.
2. The SFI Standard Audit Procedures and Qualifications (7.2) require that "the audit team shall have expertise that includes plant and wildlife ecology, silviculture, forest modeling, forest operations, and hydrology." This team is apparently lacking expertise in at least plant and wildlife ecology.
3. The audit scope appears to be just the land management portion of the company's operations.
4. The audit scope is inadequate. Objective 8, regarding procurement practices, applies to all operations where procurement of wood from sources not controlled by the entity in question takes place. An official interpretation on procurement practices has been issued by the SFI program.

Case Study 2
1. Under Objective 1, Performance Measure 1.1, Indicator 1, Polbear must have "a periodic or ongoing forest inventory"; based on the evidence, it does not. Indicator 3 of that performance measure requires "a forest inventory system and a method to calculate growth," and Indicator 4 requires "periodic updates of inventory and recalculation of planned harvests." The company has a way to calculate harvest levels but has no inventory system. A major nonconformance will likely result.
2. Clearly, their data are inadequate.
3. You should study the spreadsheet used in determining harvest levels. You should also ask questions about the source of growth data used in the calculations to determine whether they are based on empirical sources, and if not, sources that make sense for the company forests.

Case Study 3
1. Woodgood's reforestation policy may be lacking. Objective 2, Performance Measure 2.1 reads, "through artificial regeneration within two years or two planting seasons, or by planned natural regeneration methods within five years." Woodgood does some tree planting, but is the company doing this within 2 years of harvest? Field investigations would need to determine this.
2. Yes, the company has "clear criteria to judge adequate regeneration" (Performance Measure 2.1, Indicator 2).
3. Performance Measure 2.1, Indicator 3 reads, "Minimize plantings of exotic

tree species." The standard's definition of "exotic tree species" needs to be reviewed. Woodgood relies mostly on natural regeneration but also plants some sites with species that are faster growing than native species. You must determine first whether those faster-growing species are exotics. If they are, the second question is whether Woodgood is minimizing such plantings. Some judgment is required here.
4. There are at least two other issues related to long-term forest productivity that you should consider. Performance Measure 2.1, Indicator 2 discusses "acceptable species composition and stocking rates" for regeneration. Does Woodgood define "acceptable species" in its policies?

Case Study 4

1. ISO 19011 (as referenced in the SFI Audit Procedures and Qualifications introduction) provides guidance for procedures on environmental management systems audits. In ISO 19011 6.2.3, "Determining the feasibility of the audit," three factors are cited:
    - Sufficient and appropriate information for planning the audit;
    - Adequate cooperation from the auditee; and
    - Adequate time and resources.

    None of these factors are at issue with the Laguna audit. Furthermore, ISO 19011 6.3, "Conducting document review," reads, "If the documentation is found to be inadequate, the audit team leader should inform the audit client, those assigned responsibility for managing the audit programme, and the auditee. A decision should be made as to whether the audit should be continued or suspended until documentation concerns are resolved." The lead auditor could make a decision to delay the audit, based on the lack of adequate documentation, but is not required to do so by ISO 19011. Thus the audit can continue, provided that the parties to the audit understand and agree.
2. As long as the audit team does not provide consulting advice on how to implement corrective actions based on the nonconformances found, this is a third-party audit.
3. Having one or more major nonconformances means Pearmont cannot be certified "until the audit firm verifies that corrective action approved by the lead auditor has been implemented" (SFI Audit Procedures and Qualifications 6.2).

Case Study 5

1. As defined in the SFI Standard, a major nonconformance means that "One or more of the SFIS performance measures or indicators has not been addressed or has not been implemented to the extent that a systematic failure of a Program Participant's SFI system to meet an SFI objective, perform-

ance measure or indicator occurs." Objective 3, Performance Measure 3.1 requires compliance with all water quality laws and meeting or exceeding BMPs, and Indicator 4 is "Monitoring of overall BMP implementation." Water quality laws may not have been complied with, and BMPs were not monitored in this instance, but the company forester had good records on monitoring for BMP compliance. The washout appears to be an isolated incident and doesn't indicate a systematic failure. Additional questions about monitoring after major storms and a look at similar sites would be appropriate. Assuming that no other problems are found, a minor nonconformance is probably warranted.
2. You could have completed a round of road monitoring. If you had found the problem and documented it, you could have made a good case for no nonconformance, even if the culverts hadn't yet been repaired.
3. An auditee is well advised not to suggest that auditors are being unreasonable. Perhaps because it had been a long day, they were unable to put this washout in perspective.

Case Study 6

1. Yes, the lead auditor should have confirmed with the vice-president that there would be cooperation in the audit, something essential to a third-party certification audit. ISO 19011 is clear about the need for cooperation from the auditee—the company vice-president. This auditee is not cooperating, and prior to the audit the vice-president should have told the team that a critical staffer would be absent.
2. As long as other personnel can adequately provide the information the auditors need, the audit could continue. That no one can serve that role for Chenawgwa Forest Products Company, however, is a major problem for this audit. The audit really should be postponed.
3. At the opening meeting the vice-president suggested not viewing the 33,000 acres managed by the absent staffer, but the auditee should not decide what the audit team can or cannot see. Again, the audit should be postponed until the field manager can be present.

Case Study 7

1. G1 and G2 species (critically imperiled and imperiled) are cited specifically in Performance Measure 4.1, Indicator 3: "Plans to locate and protect known sites associated with viable occurrences of critically imperiled and imperiled species and communities." The company must determine whether its land harbors any G1 or G2 species and make plans to protect them if they do exist.
2. No. The forester's argument does not hold up.
3. Besides plans to locate and protect special sites, Performance Measure 4.2, Indicator 1 requires "collection of information on critically imperiled and

imperiled species and communities and other biodiversity-related data through forest inventory processes, mapping, or participation in external programs." Further requirements are found under Objective 8 (procurement): Auditees are required to provide information on G1 and G2 species to landowners.

Case Study 8

1. Formal requirements for an eligible group can be found in "Manual for Group Organizations; Group Managers and Group Members" under Standard Operating Procedures (SOP-01). Under 3.1, Legal Requirements, the group entity must be a legally registered company or organization and have an official and permanent address and an IRS tax identification number. Other requirements are outlined under SOP-01. Based on the information provided, it is likely that the consulting forester's group would be eligible for group certification, but more investigation is needed for a definitive judgment.
2. Performance Measure 3.1 of the Tree Farm Standard reads, "Forest owners must have a written forest management plan consistent with the scale of forestry operations of the property." If the combination of the consulting forester's policies and individual management plans meets the requirements as outlined in 3.1, they may be found to be in conformance.
3. Indicator 3.1.1 describes what information, at a minimum, must be included in a management plan: Title page, with basic information like the name and location of the Tree Farm, the name of the forester who wrote the plan, and the date of the plan; type of ownership; the owners' goals, which must be appropriate to the management objectives; and a tract map noting stands, conditions, and important features, such as special sites; and management recommendations that address wood and fiber production and wildlife habitat. A plan can cover other aspects of forest management at the discretion of the owner.

Case Study 9

1. The group could be eligible if it meets the requirements under SOP-01.
2. The right of first refusal creates an additional relationship that might result in a conflict of interest for the group manager, and thus it could affect eligibility.
3. SOP-01 details many requirements associated with creating a group, but an arrangement in which the group manager acquires a right of first refusal on timber from lands where it produced the management plan may cause problems. There may be ways to avoid the conflict of interest, but that is the work of the group manager and entity, not the auditor. This situation would need to be scrutinized carefully

Case Study 10

1. Performance Measure 8.1 reads, "Forest management practices must recognize historical, biological, archaeological, cultural, and geological sites of special interest." Further, Indicator 8.1.1 reads, "Management plan and forest operations identify and manage for special sites in a manner consistent with forest owner's objectives, the unique features of the site, and the size and scale of the property." Although the location was apparently chosen based on recreation objectives, the buildings also house forest management equipment, and therefore the situation is covered under the standard.
2. The language in the landowner's management plan is probably the key here. The plan says "every effort will be made to ensure that these archaeological sites will remain undisturbed," yet here is a site that might contain Native artifacts has been disturbed. The initial review probably indicates nonconformance.
3. The nonconformance is likely minor, even though inadequate care was taken: This was an isolated case and no evidence of historical use was found. An acceptable remedy or corrective action would be to require that the management plan be revised to indicate that other potential archaeological sites would be carefully examined before forest management activities were undertaken. Followup audits would verify this.

Case Study 11

1. Yes, the management plan can be changed. But if the landowner has really changed his objectives to allow for or even encourage wildlife openings, he needs to rewrite the plan to reflect that.
2. Standard 4, Reforestation, is the applicable standard. A nonconformance probably should be issued for this situation; whether it is minor or major depends on the extent of the problem.
3. Cost alone is not a sufficient reason for not conforming to a standard or the management objectives. In this situation, a sufficient remedy or corrective action might simply be to change the management plan to match the owner's new objective and to be cognizant of cost, but the plan also must make better provision for future regeneration, since the new objective would presumably not be to convert the whole Tree Farm to grass.

Case Study 12

1. Principle 5, Criterion 5.6 reads, "The rate of harvest of forest products shall not exceed levels that can be permanently sustained." Furthermore, Indicator 5.6.a. requires that the harvest rate be based on the management objectives, harvest records, and growth-and-yield estimates "as derived from stand table projections and/or published growth models." The company's method to calculate growth-and-yield is grounded in empirical

data, even though regular inventories have not been conducted. Visual inspection of the forest suggests overcutting is not occurring.
2. Allowable harvest levels can be calculated sufficiently for the time being.
3. Since the managers are already planning a full inventory to update their data, a condition (minor nonconformance) should be issued; the corrective action is what the company has proposed: a complete inventory within 5 years and recalculation of harvest levels based on the new data.

Case Study 13

1. Principle 10, Plantations, is most relevant. Criterion 10.1 states, "The management objectives of the plantation, including natural forest conservation and restoration objectives, shall be explicitly stated in the management plan, and clearly demonstrated in the implementation of the plan."
2. The FSC Standard for the US Southeast does not explicitly mandate that clearcuts be less than a particular size, although guidance language in the appendix suggests that 40 acres to 80 acres may be an appropriate maximum range that ensures economic viability of harvests while providing for other forest values.
3. A condition (minor nonconformance) is likely justified by Indicator 10.2.e, which reads, "On areas already converted to plantations, even-aged harvests lacking within-stand retention are limited to forty acres or less in size, unless a larger opening can be justified by scientifically credible analyses," because no such analysis was provided.

Case Study 14

1. FSC Principle 7 (Management Plan), Criterion 7.1 requires a written management plan and supporting documents. With turnover of personnel, the lack of a detailed management plan could disrupt management activities on the forest. The abbreviated management plan covers only some of the items listed in the standard, and it is lacking in detail.
2. The plan does not include information on rare, threatened, and endangered species management, and description and justification of harvesting techniques and equipment. There may be other missing pieces as well.
3. The lack of an adequate management plan could compromise conformance with other principles of the FSC standard, including Principles 1 (Compliance with Laws and FSC Principles), 5 (Benefits from the Forest), 6 (Environmental Impact), and 8 (Monitoring and Assessment).

Case Study 15

1. Criterion 4.5 and its indicators are applicable: "Appropriate mechanisms shall be employed for resolving grievances and for providing fair compensation in the case of loss or damage affecting the legal or customary rights, property,

resources, or livelihood of local peoples. Measures shall be undertaken to avoid such loss or damage."
2. The company should have agreed to and carried out a formal negotiation meeting with the local firm.
3. Employing local firms and cycling money from forest operations is the topic of Principle 4, "Community Relations and Workers' Rights," under which forest management operations are required to maintain or enhance the long-term social and economic well-being of forest workers and local communities." Employing foreign workers is appropriate if there is clear evidence that no local workers are available.
4. The failure of the company to even communicate with the local firm suggests that a condition is appropriate.

5 Resources

This chapter provides an overview of other information resources on forest certification, including how to get started as an auditor.

Getting started as an auditor

Anyone interested in getting started in forest certification auditing must establish his or her qualifications as an auditor (as listed in "Auditor qualifications" in Chapter 3). These qualifications can be evidenced by training, experience, or certification as an auditor, which may, in turn, be based on some combination of training, experience, recommendations, and observation of audits. Qualification requirements differ by scheme and, to some extent, by audit firm.

Meeting minimum qualifications may not be enough to find employment as an auditor, however, as there are many more qualified auditors than there are auditor openings, and all new auditors will face the age-old challenge of getting experience in a position that requires prior experience. To find employment as a forest auditor, therefore, new entrants should plan to go beyond minimum requirements, network with old and new acquaintances to find work, and be willing to take training assignments without compensation to gain experience.

Candidates with previous experience developing, operating, and internally auditing forest certification schemes or parts of schemes (e.g., water quality best management practices) have an advantage in seeking employment as certification auditors. In addition, although most audit firms offer training in their systems, prior knowledge in the forest certification standards and training in auditing provide a more solid basis for firm-specific, on-the-job training. To this end, a week-long EMS lead auditor course is excellent training for any auditor and also may provide contacts. And of course, the one-day workshop on which this book is based is a good way to get more of this scheme-independent overview of auditing principles as they are applied to forestry.

In addition to prior preparation, a familiarity with and willingness to audit to a variety of standards provides more possibilities for openings, especially for combined-standard audits. This is particularly true as markets for certification to existing standards mature and as new standards emerge, such as for Tree Farm Group or Master Logger certification.

RESOURCES

Forest certification websites

All forest certification schemes maintain websites where potential auditors and auditees can find out more about those schemes and, often, about audit firms. The following list provides a fairly complete starting point for North America.

Environmental management systems
ISO EMS: http://www.iso.org/iso/en/iso9000-14000/index.html
ISO EMS development: http://www.tc207.org/

Forest certification systems
CSA: http://www.csa-international.org/product_areas/forest_products_marking/
FSC International: http://www.fsc.org/en/
FSC Canada: http://www.fsccanada.org/
FSC US: http://www.fscus.org/
Green Tag: http://www.greentag.org/
SFI: http://www.sfiprogram.org/
Tree Farm: http://www.treefarmsystem.org/
PEFC International: http://www.pefc.org/
PEFC Canada: http://www.pefccanada.org

Auditor certification
Canadian Environmental Auditing Association:
   http://www.ceaa-acve.ca/certify.htm#msapro
RABQSA: http://www.rabqsa.com/cp_com.html